LIGHTING UP THE SKY

NORTHERN LIGHTS

SETH KINGSTON

PowerKiDS press
New York

Published in 2021 by The Rosen Publishing Group, Inc.
29 East 21st Street, New York, NY 10010

First Edition

Portions of this work were originally authored by Kristen Rajczak and published as *The Northern Lights*. All new material in this edition authored by Seth Kingston.

Editor: Elizabeth Krajnik
Book Design: Reann Nye

Photo Credits: Cover krissanapongw/iStock/Getty Images Plus/Getty Images; series art Zoteva/Shutterstock.com; series art Abstractor/Shutterstock.com; p. 5 Marc Marchal/500px Prime/Getty Images; p. 6 V. Grazyna Wilk-Scott/500px/Getty Images; p. 7 Mondadori Portfolio/Getty Images; p. 8 muratart/Shutterstock.com; pp. 9, 14 Simon's passion 4 Travel/Shutterstock.com; p. 11 VW Pics/Universal Images Group/Getty Images; p. 13 Courtsey of NASA; p. 15 MikeDrago.cz/Shutterstock.com; p. 16 Petri jauhiainen/Shutterstock.com; p. 17 NASA/Getty Images News/Getty Images; p. 19 © iStockphoto.com/chemc; p. 21 (top) MAGNIFIER/Shutterstock.com; p. 21 (bottom) Denis Belitsky/Shutterstock.com; p. 22 Noppawat Tom Charoensinphon/Moment/Getty Images.

Cataloging-in-Publication Data

Names: Kingston, Seth.
Title: Northern lights / Seth Kingston.
Description: New York : PowerKids Press, 2021. | Series: Lighting up the sky| Includes glossary and index.
Identifiers: ISBN 9781725318618 (pbk.) | ISBN 9781725318632 (library bound) | ISBN 9781725318625 (6 pack)
Subjects: LCSH: Auroras--Juvenile literature.
Classification: LCC QC971.4 K56 2021 | DDC 538'.768--dc23

Manufactured in the United States of America

Some of the images in this book illustrate individuals who are models. The depictions do not imply actual situations or events.

CPSIA Compliance Information: Batch #CSPK20. For Further Information contact Rosen Publishing, New York, New York at 1-800-237-9932.

CONTENTS

LOOK UP!

Imagine you're in Alaska. It's December and you're outside getting wood to keep the fire going inside the fireplace. You look up and see yellow, green, blue, and purple lights dancing in the sky. This phenomenon, or event, is called an aurora. Auroras were named after Aurora, the Roman goddess of **dawn**.

The most famous auroras are called the aurora borealis, or northern lights. "Borealis" comes from the Greek god of the north wind, Boreas. Aurora borealis means "northern dawn" in Latin. Read on to learn more about the northern lights!

The northern lights have fascinated, or greatly interested, people for thousands of years. It's easy to see why!

SCIENTIFIC DISCOVERIES

The northern lights first got the name aurora borealis around 1620 when scientists Galileo Galilei and Pierre Gassendi wrote about them. Galileo thought auroras came from sunlight reflecting in Earth's **atmosphere**.

SEEING THE LIGHT

As early as 2010, people in Canada began seeing a new type of aurora, which they named STEVE. STEVE is a dim purple and green ribbon in the sky. However, STEVE isn't actually an aurora.

Around 1895, Kristian Birkeland, shown here, discovered that **electrons** meeting Earth's magnetic field cause auroras.

In 1790, Henry Cavendish discovered that the auroral light forms about 62 miles (100 km) above Earth's surface. In 1903, Kristian Birkeland discovered that electricity flowing through the gas of Earth's upper atmosphere causes the light we see as auroras. In the 1950s, scientists used special rockets to collect knowledge about auroras. Today, we know that charged **particles** from the sun cause auroras.

WHAT CAUSES AURORAS?

The sun gives off light, heat, and small particles that travel in solar wind, which are winds the sun produces. Solar winds reach Earth in about 40 hours, where they meet Earth's magnetic field and atmosphere. Earth's invisible magnetic field pushes some of the energy and particles away. However, the magnetic field pushes some energy and particles toward Earth's poles.

Earth's magnetic field creates a shield to keep harmful particles from reaching Earth's surface. However, solar winds reach Earth's atmosphere at the poles to create beautiful auroras.

When particles reach the poles, they meet oxygen and nitrogen **atoms** in Earth's upper atmosphere. Electrons and **protons** in the solar wind knock away electrons from the oxygen and nitrogen atoms, leaving behind energized **ions**, which give off light energy we see as colorful auroras.

COLORED LIGHT

The most common color of aurora is yellow-green. However, auroras can be red, pink, blue, or purple. The aurora's color depends on which gas atoms the particles from the solar wind meet and how high in the atmosphere the particles and atoms meet.

The yellow-green color most people think of when they imagine an aurora is due to oxygen atoms about 60 miles (97 km) above Earth's surface. However, oxygen atoms at about 200 miles (322 km) above Earth's surface produce red auroras. Nitrogen causes blue or purple lights. Oxygen and nitrogen also give off ultraviolet light, which can only be seen with special cameras on spacecraft.

SEEING THE LIGHT

An aurora's color also depends on what type of collision, or crash, occurs between the particles in the solar wind and the atoms or **molecules** in the atmosphere. Atomic nitrogen produces blue, while molecular nitrogen produces purple.

This aurora over Grasslands National Park in Saskatchewan, Canada, has many colors. Which atoms or molecules do you think produced these colors?

CURTAINS AND CROWNS

Auroras come in many different shapes. In fact, no two auroras look alike. Depending on where the observer is standing, an aurora may look like a curtain or form a corona, or crown. Some auroras seem to be static, or fixed, in the sky, while others move like dancing ribbons. Over time, auroras can change shape and brightness.

Auroras form ovals above Earth's magnetic poles. The parts of Earth under these ovals are called the auroral zones. Auroral zones move toward the poles during low solar activity and reach farther toward the equator during periods of high solar activity. However, no one standing on Earth can see the full auroral zone.

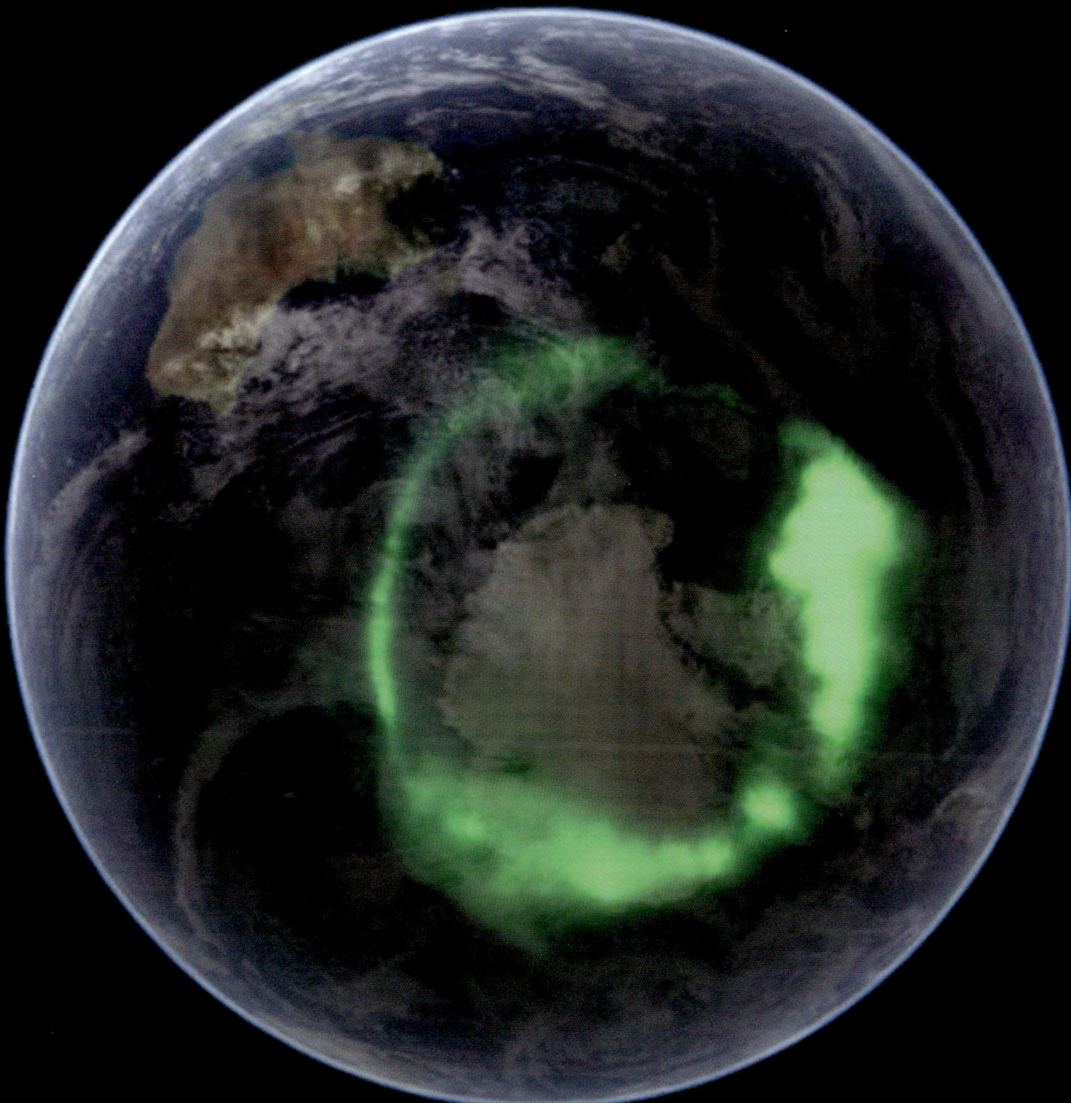

Entire auroral ovals can only be seen from space. This image shows the auroral oval over the South Pole.

WHERE AND WHEN TO WATCH

The auroral zone in the Northern **Hemisphere** exists over the northern parts of Scandinavia and Iceland, the southern part of Greenland and the Hudson Bay in Canada, central Alaska, and the coast of Siberia. However, these places aren't always easy to get to.

SEEING THE LIGHT

Every 11 years, the sun's magnetic field goes through a **cycle** called the solar cycle. At the middle of the cycle, or solar maximum, activity on the surface of the sun is at its highest. Auroral activity increases during solar maximums.

Some of the best times to watch auroras are near the equinoxes in March and September and on moonless nights.

Luckily for observers, the northern lights reach farther south during periods of high solar activity. When the sun's activity increases, the northern lights also become brighter and more colorful.

The northern lights exist year round. However, they're best seen on nights during the winter months because the air is cleaner and there's less light pollution. Nights are also longer during these months.

THE GREAT SOLAR STORM OF 1989

On March 10, 1989, the sun experienced a major solar flare, which sent particles toward Earth at a rate of 1 million miles (1,609,344 km) per hour. These particles hit Earth's magnetic field on March 12, creating beautiful northern lights that lasted for two days. The particles from this storm pushed the auroral zone southward to Florida and Cuba.

The solar flare from March 10, 1989, had a strength of X8. This image of a solar flare from October 28, 2003, had a strength of X18. The solar storms this solar flare produced were some of the most powerful ever recorded.

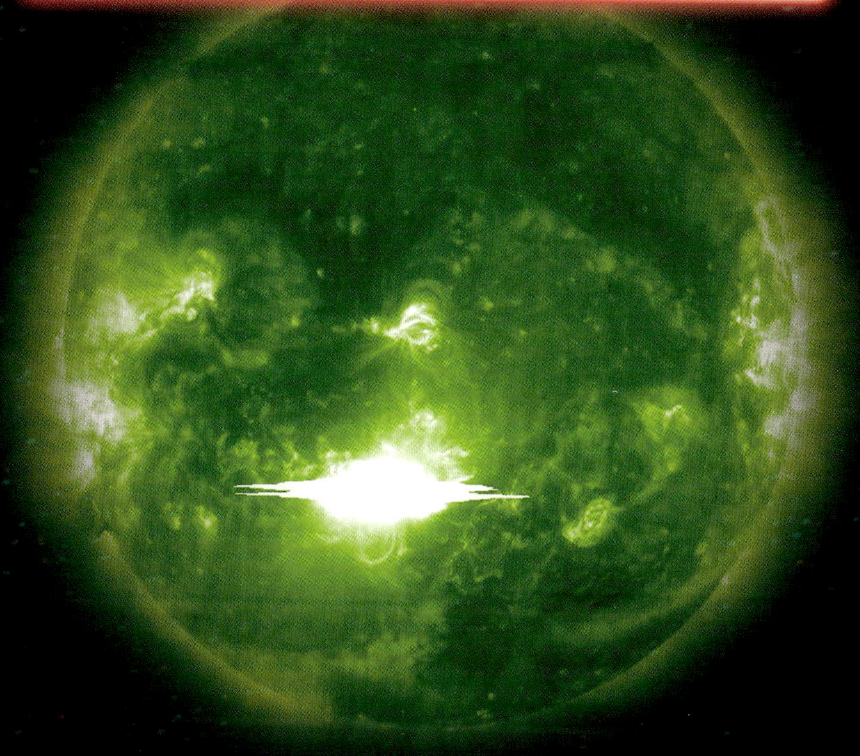

The solar flare also caused a geomagnetic storm, which was so strong that it caused people throughout the entire province of Quebec, Canada, to be without electricity for 12 hours. Other places had problems with their power too.

AURORA AUSTRALIS

In the Southern Hemisphere, auroras are called aurora australis, or southern lights. *Australis* is Latin for "of the south." The same solar winds and collision of particles that cause the northern lights also cause the southern lights. In fact, the southern lights often happen at the same time and with shapes and colors similar to the northern lights.

However, fewer people have the chance to see the southern lights than the northern lights. The southern lights commonly take place over the Indian Ocean and Antarctica. They can be seen from Tasmania, New Zealand, and even parts of southern Chile, Argentina, and South Africa.

SEEING THE LIGHT

One of the best places to see the aurora australis is South Georgia Island. However, the absolute best place to see the southern lights is Antarctica because the lights are brighter there.

The southern lights often have more pinks, purples, oranges, and golds than the northern lights.

TIME TO SHINE!

Now that you've learned all about auroras, it's your turn to see one in real life. The best place to watch the northern lights is as close to the North Pole as you can get. There, the lights will be at their brightest. Plan to visit this area between September and March for the best light show.

To best see the beautiful lights, it must be very dark, the weather must be calm, and the skies clear. Late night and early morning are the best viewing times. Watch the sky carefully. Auroras don't last very long!

Many people take vacations to places such as Iceland just to see the northern lights.

BELIEFS ABOUT AURORAS

Today, we know the science behind auroras. However, just a few hundred years ago, people had no idea what caused them. Different groups of people had different beliefs about auroras. The ancient Greeks believed Aurora was the sister of the sun and moon. The ancient Romans believed the northern lights were Aurora, the goddess of dawn.

Some groups of Inuit people in North America believed auroras were the spirits of dead people playing a ball game. In Finland, people believed that auroras were created by the mythical firefox running through the snow so fast that sparks would fly into the night sky.

GLOSSARY

atmosphere: The whole mass of air surrounding Earth.

atom: The smallest part of a substance that can exist alone or in combination.

cycle: A repeating period of time.

dawn: The time of day when sunlight first begins to appear.

electron: A very small particle that has a negative charge of electricity and travels around the nucleus of an atom.

hemisphere: One-half of Earth.

ion: An atom or group of atoms that carries an electric charge.

molecule: The smallest possible amount of something that has all the characteristics of that thing.

particle: Any one of the very small parts of matter, such as a molecule, atom, or electron.

proton: A very small particle that exists in the nucleus of every atom and has a positive charge of electricity.

INDEX

WEBSITES

Due to the changing nature of Internet links, PowerKids Press has developed an online list of websites related to the subject of this book. This site is updated regularly. Please use this link to access the list:
www.powerkidslinks.com/lus/northernlights